FORD TRACTORS

Jonathan Whitlam

First published 2018

Amberley Publishing
The Hill, Stroud,
Gloucestershire, GL5 4EP

www.amberley-books.com

Copyright © Jonathan Whitlam 2018

The right of Jonathan Whitlam to be identified as the Author
of this work has been asserted in accordance with the
Copyrights, Designs and Patents Act 1988.

All rights reserved. No part of this book may be reprinted
or reproduced or utilised in any form or by any electronic,
mechanical or other means, now known or hereafter invented,
including photocopying and recording, or in any information
storage or retrieval system, without the permission in writing
from the Publishers.

ISBN: 978 1 4456 7765 1 (print)
ISBN: 978 1 4456 7766 8 (ebook)

British Library Cataloguing in Publication Data.
A catalogue record for this book is available from the British Library.

Typeset in 10pt on 13pt Celeste.
Typesetting by Amberley Publishing.
Printed in the UK.

EU GPSR Authorised Representative
Appointed EU Representative: Easy Access System Europe Oü, 16879218
Address: Mustamäe tee 50, 10621, Tallinn, Estonia
Contact Details: gpsr.requests@easproject.com, +358 40 500 3575

Contents

	Introduction	5
Chapter 1	Built to Save a Country	7
Chapter 2	The N Dynasty	12
Chapter 3	The Ferguson Connection	21
Chapter 4	Austerity Britain Gets The Austerity Major	27
Chapter 5	Major Step Forward	31
Chapter 6	Built For The World	42
Chapter 7	Perfect 10	62
Chapter 8	Futuristic Features	80
Chapter 9	The Name Disappears	88
Chapter 10	Legacy	94

Introduction

Henry Ford is one of the best-known industrialists of all time and is credited with changing the world forever, thanks to the use of his mass production techniques to build the Model T car and sell it cheaply enough to be affordable to the masses. But this was not all, because Henry Ford also did much the same for agriculture in manufacturing the first successfully mass-produced tractor.

Coming from a farm, Henry realised just how hard life working on a farm actually was before the advent of mechanisation. Even with the advent of steam power on the farm, most of the work was still done by hand, and the young Henry had first-hand experience of the drudgery of it all. Therefore, even when he was working on his first car prototypes,

A preserved Ford Model T car. This was the first mass-produced automobile and was a huge hit due to its affordable price.

he was also putting ideas into place to build a small tractor, a machine that was light and versatile enough to be used by even the smallest farmstead.

Cars had to take first priority and even with the success of the Model T his fellow investors would not allow spending money on a tractor. Therefore Henry started his own company, separate to the Ford concern, called Henry Ford & Son, which only had himself, his wife and his son as shareholders. It was under the auspices of this company that the first tractors were built, but circumstances soon allowed Henry to take a major shareholding in the original Ford Motor Company and bring it under family control, thereby uniting the car and tractor businesses.

The tractor that Ford eventually put into production was ahead of its time, used unitary construction principles and was built in huge numbers the like of which had never been seen before. It would change the way tractors were built and farmers farmed – it was indeed the first 'modern' tractor!

Chapter 1

Built to Save a Country

The First World War was still raging in 1917 and German U-Boats were sinking the shipping bringing essential food supplies to Great Britain to such an extent that the country was facing starvation at a crucial time of the war. Things had been made worse by the fact that British agriculture had been badly neglected and there were thousands of acres not even under the plough. The government, under the auspices of the Ministry of Munitions, realised that the only answer was mechanisation on a scale the country's farmers had never seen before, just to get as many acres back into cultivation as possible. To do that it was obvious that tractors would be needed, even though they were still very much a new technology. Early British tractor manufacturers were not in a position to meet the huge demand and so it was to the USA that the Ministry looked; after all, it was in America that the tractor had been born.

The result was many tractors coming into the country from the United States, including the International Harvester Titan 10-20 and Mogul 8-16, but these were not ideal for British farms and the Ministry of Munitions was looking for something a bit more versatile. This was when the head of the British arm of the Ford Motor Company, Lord Perry, came in. Knowing of the Ford tractor prototypes on test in the USA, he arranged for two to be imported into Britain for trials in January 1917.

These prototypes were very advanced machines and many were built for testing, being called the X Series. Designed by Eugene Farkas, these early machines were pretty much what would eventually go into production, except for a few tweaks here and there, and were powered by a 20 hp Hercules engine. Their real secret, though, was that they were built using the unitary construction method, which meant that the sump of the engine and the transmission formed the 'backbone' of the tractor instead of everything being mounted separately on a heavy chassis, as was the norm for almost all other tractors then being built. This concept had been first used successfully on the Wallis tractor, but was not widely accepted without Ford's mass-production techniques, even when the design was later taken over by Massey-Harris.

A thermo-syphon type cooling system was fitted to these tractors, which incorporated an air washer, while the gearbox provided three forward gears and a single reverse. Brakes were not fitted.

This early restored MOM tractor from 1917 has received the lettering along the sides of the fuel tank that was fitted to the first X Series prototype to arrive in Britain in January 1917. That first tractor was painted white but this example has been painted the same green as was discovered on an older MOM tractor belonging to the same Norfolk owner.

The first two tractors brought over to the UK soon proved their worth and led to the Ministry of Munitions placing an order with Henry Ford to produce 6,000 of them. The very first one was painted white and had the words 'Peace, Industry, Prosperity' written in large script down each side of the fuel tank. It was originally intended to build these tractors in Britain, but the war made this impractical, and so it was decided to build them in Dearborn, Michigan, in the USA.

There was a lot of work to do to get the factory production line up and running and only 254 tractors would be built in 1917, but this would be substantially increased the following year. Tractors began to arrive in the UK during October 1917, and because these early tractors were brought over to Britain and remained under the ownership of the Ministry of Munitions, they quickly became known as the MOM tractors, because so far no name had yet been decided on for them, so quickly had they been rushed into full production.

By the spring of 1918, production was running so high that the home market could begin to be catered for. At the same time the tractor was finally given a name – the Fordson Model F.

To start with, the Model F was very similar to the MOM tractors but gradually changes were introduced, including solid sides to the radiator instead of the holes that featured previously, these disappearing completely by 1919. Production was moved from its original site in Dearborn in 1921, with the tractors now being built in the new, massive Rouge River factory. The Hercules engine was also dropped in favour of a very similar Ford-built unit, which was still of four cylinders and with an output of around 20 hp.

Although fitted with a new fuel tank and other items to replace those that have decayed over the last century, this MOM has been only lightly restored and is believed to be one of the earliest in Britain.

The Hercules four-cylinder side-valve engine produced around 20 hp and was fitted with a Holley manifold, as can be seen clearly here.

A wooden steering wheel was provided, this one being a reproduction. The small toolbox underneath is original, as is the gear lever.

Left: An early Fordson Model F from 1918 showing the basic build of the tractor and clearly demonstrating the unitary construction methods employed.

Below: The Fordson name was applied first in the spring of 1918 to Model F tractors, which were sold first in the USA. The name has its origins in the company set up to build the new tractor: Henry Ford & Son.

A later Model F from 1923, fitted with a Ford-built engine with a modified electrical system.

Long rear mudguards were also introduced in the early 1920s to counteract the rather terrifying tendency of the Fordson to rear up backwards if its implement encountered something solid. The result was that the driver could be fatally crushed to death and so the long mudguards, complete with small toolboxes in the ends, grounded first and prevented the tractor from toppling backwards.

In 1919 Fordson F production also started in a Ford factory in Cork, Ireland, but a worldwide depression in agriculture saw production here finish, and even the Rouge River plant ceased production of the Model F in 1928.

During its life, the first Fordson had achieved much. Not only had it been a huge success in America, but it had also helped save Britain from being forced into submission under pressure from the German navy, all while bringing mechanised farming to thousands of farms the world over.

Chapter 2

The N Dynasty

In 1929 production of the Model N tractor, the replacement for the Model F, began in Cork, Ireland. This was a very similar machine to the Model F but featured an extra 3 hp after the engine was redesigned with an increased bore and new cylinder head. Other improvements included heavier front axles and cast front wheels that helped to keep the front of the tractor firmly planted on the ground, while the long rear mudguards remained a feature of the Irish machines.

It had always been the plan to move tractor production to Dagenham in Essex and so, in 1932, the production line tooling was gradually moved to England. Then, in 1933, following a limited run of tractors using both Irish and English components, the Fordson N appeared in a new guise.

The Model N was first built in Cork, Ireland, from 1929.

As with the last Model F tractors, the Model N was fitted with the long rear mudguards, complete with two small toolboxes incorporated into the ends.

It certainly looked different, with a new dark blue paint job combined with red wheels. Various other improvements to the cooling system saw the use of a water-washer air cleaner, which very cleverly used a hollow steering wheel as an air intake.

Dagenham was a huge purpose-built factory built on Essex marshland that not only included its own dock, but also railway sidings. In fact, this huge facility basically took raw materials in at one end and spat finished cars, trucks and tractors out of the other!

British farmers took to the new Fordson Model N and it sold well with quite a few exported, especially when a row crop version with a tricycle type format was built for sale in the USA. After the initial birth of the Model F and the MOM tractors being imported into the UK from the United States, now the position had been reversed to some extent, with tractors from Essex making their way over to the USA.

In 1937 a new, brighter orange colour scheme was introduced to make the Fordson look fresher, although in fact changes had been minimal, with the engine now being equipped with a higher compression head to produce a little more power. Unfortunately, problems arose from this, as the tractor required proper high-grade TVO to run on, and many farmers just made do with inferior products that resulted in engine problems.

The Second World War was looming, and when Britain declared war on Germany in 1939, Ford already had a large stockpile of tractors ready to be sent out to increase food production once again. The result was that Fordson tractors and ploughs could soon be seen turning over every available inch of spare ground, including parkland and roadsides.

When production of the Model N moved from Cork to Dagenham in Essex, several small changes were made to the design, as well as a new blue and red colour scheme.

Steel wheels were the normal specification for the Model N and the cleats on the rear wheels helped gain traction. With a heavier front axle, the long rear mudguards were deleted from the specification.

Above: Pneumatic tyres gradually became an option when supplies of rubber allowed and certainly improved the ride quality of the tractor.

Right: The view from the seat of a blue Fordson N, showing the air cleaner intake and toolbox.

Left: The Fordson All-Around was a row crop version of the Model N which was produced for export to the USA. This one has a V-twin front wheel arrangement.

Below: From 1937 a brighter all-over orange colour was used in an attempt to refresh the Model N and also announce more power being available.

The orange Model N featured a new high compression head but this caused some reliability issues if low-grade fuel was used.

The row crop version was still built and this one is equipped with a single front wheel, making it a true tricycle machine.

Because of the bright rows of orange tractors lined up at Dagenham, it was decided to change the colour from orange to green to make it harder to spot them from the air. The green tractors then remained pretty much unchanged until shortages of raw materials began to bite and, in 1942, much narrower rear mudguards were fitted to save on metal. At least tractors were still being built, the Dagenham factory being the only one still allowed to produce tractors for agricultural use in large numbers, and even then many were also used on airfields the length and breadth of the country for mowing grass and pulling airplanes and ancillary equipment.

The Model N was an undoubted success but it was in many ways outdated when it was first built in Cork in 1929. After all, this was a design that remained only slightly changed from the original MOM tractors of 1917! Production of the Model N continued up to 1945 despite a brand-new, ultra-modern tractor design being put into production by Ford in the USA from 1939.

All-over green was adopted in 1940 due to the orange tractors being vulnerable to German air raids. Most new tractors at the time would have still been fitted with steel wheels rather than pneumatics, which were often added later.

Above: Showing its working life well, this Fordson N takes a break while cultivating.

Right: Narrower rear wings were implemented in 1942 to save on steel due to material shortages.

Several Fordson N tractors have been retrofitted with a diesel engine after the original unit wore out. This example, using a Perkins engine, is a very neat conversion.

Chapter 3

The Ferguson Connection

Harry Ferguson was a very clever inventor from Ireland who had developed a mounted plough for the Fordson F and then went on to design his Ferguson System of hydraulic draft control and three-point linkage, which was designed into a new tractor called the Ferguson Type A in 1936. David Brown Limited of Huddersfield, famous as gear manufacturers, built this for Ferguson.

The system that Ferguson invented not only allowed fingertip control of a mounted implement so that it could be lifted and lowered out of work instead of just pulled behind. It also allowed for hydraulic control of the draft, adjusting the height of the implement in the ground to follow contours and, if an immoveable object was encountered, the clever geometry of the three-point linkage meant that there was no rearing backwards with the front of the tractor in the air; instead, the forces acted through the top link and kept the tractor's nose firmly on the ground. All very clever stuff and a design that allowed a light, small tractor to operate extremely efficiently and without any added ballast weight.

There was a problem though. To get the best out of the Ferguson System, a whole set of matched implements had to be bought to go with it, making it a very expensive tractor to buy, and farmers in the UK were reluctant to spend the money. This led to friction between Ferguson and Brown, the latter having large numbers of tractors collecting in the factory yard. Brown suggested changes to the tractor to make it more affordable to produce, but Ferguson would have none of it and soon travelled across the Atlantic to meet with Henry Ford.

Ferguson was an expert at demonstrating his creation and the two Ferguson Type A machines he took with him were soon working on the Ford farms, giving a demonstration to Henry Ford. Ford was very impressed with the little tractors and he and Ferguson made a handshake agreement there and then to produce a tractor incorporating the Ferguson System.

In a remarkably fast time the new tractor was ready. This was something very different to what had gone before and used sleek modern styling, a modern petrol engine and, of course, the Ferguson System with three-point linkage. Launched in 1939, the 9N, as it was called, was a revelation to American farmers as it could do what previously had needed a much larger and heavier machine to do – and what's more, the 23 hp 9N could often do it better!

The Ford 9N looked unlike any other tractor before it and was a landmark machine in the history of farm mechanisation.

The Ford and Ferguson names were united on two separate badges fitted to the front radiator of the Ford 9N.

Automotive style gauges were used on the 9N, giving a much more modern driving experience than the Model F or N.

The secret of the success of the 9N when in work was the Ferguson System of three-point linkage and hydraulic draft control, which required a range of matched implements, such as this two-furrow plough, to be bought with the tractor.

Ferguson took over the sales of the new tractor and its matched range of implements while Ford's team designed the tractor and then put it into production. Ferguson's prowess at demonstrating the tractor was no doubt the main reason for its startling success, and the fact that Ford's mass-production facilities were put behind it also meant that it was affordable.

In 1942, due to material shortages, the 2N was introduced with reduced features as an austerity model to keep production going, but as the situation eased the 2N was soon back to pre-war specification.

But Ferguson was not happy for long. He wanted the 9N to be put into production at Dagenham, in Essex, to replace the Model N. Many thousands of 9N and 2N tractors were exported to the UK when a TVO version was built as part of the Lend-Lease arrangements. But when it came to replacing the Model N the British division of the Ford Motor Company could not justify the cost and time of re-tooling to produce the new tractor and went down a stop-gap route instead with the Major. This incensed Ferguson and resulted in him starting to build his own tractor in Coventry from 1946. Thus, the Ferguson TE20 was born.

Back in the States, Ford replaced the 2N with the 8N, but Ferguson took the firm to court over infringement of patents as this tractor used the Ferguson System hydraulics. It took years to sort out the case through the courts, which Ferguson eventually won, but small tweaks to the hydraulic system allowed Ford to carry on with production and the 8N soon evolved into a whole series of tractors that were produced in America up to 1964, and included the NAA Golden Jubilee of 1953, which was the first to feature the new Ford Red Tiger engine. This model formed the basis for the Workmaster and Powermaster ranges of 1955, which in turn became the 2000 and 4000 series in the early 1960s, with a whole host of varying features including row crop models, powershift transmission and various different fuel choices.

The Ford 8N was produced by Ford themselves after splitting with Harry Ferguson and caused Ferguson to take Ford to court over its use of his patents.

The Ford NAA, or Golden Jubilee, was the next development to emerge from the 9N design.

Launched in the same year as the fiftieth anniversary of the Ford Motor Company, the NAA was fitted with this special badge as part of its front radiator cowling.

The NAA was the first Ford tractor to feature an overhead valve engine of 30 hp, which was called the Red Tiger. A four-speed transmission was fitted, as well as the option of a live power take-off.

By 1958 the range of American Ford tractors had expanded and evolved to include the Workmaster 641, its Red Tiger engine producing nearly 31 hp.

Above: The larger Powermaster 851 was of just over 43 hp and this one is in the later red and grey livery.

Left: A diesel engine resides under the colourful exterior of this Ford 961, not a common fitment as petrol was cheap and plentiful in the USA. Note the wide tread and high clearance of this tractor for row crop work.

Chapter 4

Austerity Britain Gets The Austerity Major

The reason that a similar tractor to the 9N could not be produced in Britain was simple – there just were not the materials available after the devastation of the Second World War to do it! There was no doubt that the Model N needed replacing with something more modern, and prototypes had already been built, but there was just not the ability to put anything radically new into production.

Production began in 1945 of a tractor that incorporated a new back end and transmission with the old side-valve, four-cylinder Ford engine used in the Model N. The result, the Major, could be built quite easily and in large numbers, and therefore was the answer to the problem.

This is why the old Ford engine remained in production for so long, albeit now mated to a modern transmission. The Major also stood taller and the colour had reverted back to the dark blue and red of the 1933 Model N. To the farmers who were desperate to replace worn out tractors and also to replace horses, the Major was ideal, and it was available in quantity.

The Major was also known as the E27N internally, referring to the fact it was built in England, had a rating of around 27 hp and was a tractor, the N being Ford's code to denote a farm tractor. When it came to design it was clearly similar to the Model N, with the radiator mounted at the front but now with the word 'Major' cast into the front, and a dual fuel tank positioned above the engine itself, just like on the earlier Ford tractors. The differences with the Ford 9N could not be more marked!

Ford had a success on their hands when they fitted the Major with a six-cylinder Perkins diesel engine in 1948. Diesel was becoming a popular alternative to petrol and TVO in Britain when it was realised that it offered greater efficiency and more pulling power. Perkins, based in Peterborough, had pioneered the fitting of diesel engines into the Model N as replacement engines, and the P6 proved to be a good match for the Major. With 47 hp now produced, this nearly doubled the power output.

The Major carried on up to 1951, when, finally, a brand-new tractor was ready for production. It had been a stop-gap machine, there is no doubt, but the Major had also been a success and had been instrumental in the country recovering from the stringencies of a world war.

An early Fordson Major that was original on steel wheels and is also equipped with a Miller lift attachment, hence the tall lever above the bonnet. This lift allowed for the rear cultivator to be lifted in and out of work on the headland but required a great deal of physical effort to operate.

The Major was designed to start on petrol and then switch over to TVO, and was fitted with the same engine as the Fordson N, complete with a dual-compartment fuel tank. This one is using a Massey-Harris binder in Norfolk.

The view from the seat of the Major, showing the basic controls and the similarity to the Model N. No car type gauges and instruments here!

For most farms the Major was the main tractor and as such was responsible for many duties including ploughing. This example is fitted with retractable wheel strakes on the rear wheel, which could be engaged to aid traction in sticky conditions and help the tractor maintain traction.

The Perkins P6 diesel sat neatly under the fuel tank and mated to the new rear transmission of the Major perfectly. Even with a substantial increase in power, the new back end coped easily.

When the Perkins P6 diesel engine became available as a factory-fitted option, the Fordson Major really became a useful piece of equipment, with more horsepower and more torque.

This Perkins-powered Fordson Major has been updated quite a bit over the years, including a more comfortable seat and larger diameter rear wheels. It is shown drilling wheat with a Smythe drill in Norfolk.

Chapter 5

Major Step Forward

Such was the success of the Major that when a new model was launched in 1951, it still retained the same name and was known as the New Major. And this *was* new. Although using the transmission proven in the original Major, the new tractor incorporated a brand-new overhead-valve, four-cylinder engine that could be run on petrol, TVO or diesel, with power outputs from 35 to 40 hp, and also featured brand-new styling and tinwork plus a six forward and two reverse gearbox. A lighter blue colour – known as Empire Blue – was also adopted, making the new machine stand out from what had gone before.

Internally the New Major was known as the E1A, with a combination of letters designating which engine was fitted dependent on fuelling. Despite the popularity of TVO, it was the diesel version that soon became the most popular, soon giving rise to the Diesel Major name, which it eventually carried on the sides of its bonnet.

An early example of the New Major, showing off its much more modern lines that encapsulated the new engine design that could be configured to run on either petrol, TVO or diesel.

It was the diesel version that proved to be the one everybody wanted and soon the Diesel Major gained its own identity, complete with a special badge on the bonnet sides.

The engine fitted to the New Major tractors, in this case a diesel variant, showing the Simms injector pump mounted to the side.

Controls on the New Major were much more modern than those used on the original Major, and included useful gauges as well.

The look of the new tractor was superb and fitted in with the latest in vehicle design concepts, with not only sleek looks but also practicality, with liftable flaps to gain access to the engine, which was made possible by the fuel tank being repositioned backwards to in front of the driver. Still built along the original unit construction principles, a strengthening chassis frame was also added to the bottom of each side of the engine block.

To say the New Major was a success is an understatement! The engines, particularly the diesel version, were extremely well designed and can truly be called 'bomb-proof'. Various upgrades saw extra features added to the Major over time, including live hydraulics.

In 1957 a smaller tractor joined the Major in an attempt to cash in on the market begun by the Ferguson TE20. The 32 hp Dexta was actually based on the American Ford tractor line, but with a new three-cylinder diesel engine that was a joint venture between Ford and Perkins. A petrol version was also built in small numbers. The Dexta was aimed squarely at the smaller farm that did not require the power of the larger Major and now, with two models, Ford could offer farmers something for every requirement.

The Diesel Major was updated into the 52 hp Power Major in 1958 with more power and a new badge to show it, but it was only in production until 1960, when the Super Major replaced it. The 40 hp Super Dexta also joined the line-up, giving three models to choose from: the Dexta, Super Dexta or Super Major.

In 1963 the last incarnation of a Dagenham-produced Fordson tractor arrived in the shape of the New Performance range, with a new colour scheme including the more extensive use of white. Otherwise, these were still the Dexta, Super Dexta and Super Major we knew before, but with the two larger tractors having slightly increased power available.

Production would end in 1964, however, to make way for a brand-new Ford tractor line to be built globally. At the same time Dagenham would cease to build tractors for the first time since 1933 and the Fordson name would also disappear forever.

The Fordson Dexta arrived in 1957 to give the Major a new workmate and was aimed at those who were using Ferguson tractors in particular.

This later version of the Dexta features headlights built into the front grille rather than on stalks, and is pulling an Albion binder through a wheat crop in Norfolk.

The three-point linkage on the Dexta was also fitted with draft control, called Qualitrol by Ford.

1958 saw the replacement of the Diesel Major by the Power Major and this very original example is shown ploughing in Lincolnshire.

This Power Major is equipped with a period weather cab built by Winsam, which, although it protected the driver from the elements, must have severely restricted visibility!

Launched in response to more powerful three-cylinder Massey Ferguson models, the Super Dexta saw the original Dexta engine tweaked to provide more output.

The higher bonnet line of the Super Dexta was used to incorporate a new badge into the front end.

Ploughing with the Super Dexta and having plenty of horses to spare with 40 hp under the bonnet.

The Super Major of 1960 saw the introduction of a much more refined version of the E1A design, with not only more power, but also a long list of extra options.

New badges were produced to make the Super Major stand out from its predecessors, while the engine was still based on that original design from 1951.

With rear wheel weights fitted, this Super Major is turning over the stubble in a field in Lincolnshire.

The New Performance line-up of 1963 was the last flowering of the Fordson tractor dynasty. This was the New Performance Dexta, the smallest, which remained unchanged from its earlier version except for the new livery.

Next up was the New Performance Super Dexta, now with nearly 5 hp more than previously.

Above: Top of the range was the now more powerful New Performance Super Major.

Left: The driving platform of the New Performance Super Major was a nice place to be, with everything laid out neatly and to hand. Note the fuel tank under the steering wheel – a feature of the E1A design right from the start.

American farmers were gradually moving over to diesel power and so the Super Major was exported to fill the gap. These tractors received white bonnets and the addition of a Ford 5000 decal.

Chapter 6

Built For The World

Up until the end of 1964, Ford tractor production had been based around two very different lines in the USA and the UK. In Britain the Fordson line had remained, evolving from that original MOM tractor that arrived in January 1917. In America, meanwhile, a whole range of tractors had been developed from the 9N and were built in a variety of different versions, including row crop models. An advanced transmission had also appeared called Select-O-Speed, which was basically a very early semi-powershift gearbox. A big 76 hp six-cylinder tractor called the 6000 had also been launched, but had suffered from a few mechanical reliability issues.

The 6000 was Ford's attempt to break into the higher power six-cylinder bracket in the USA. Unfortunately, its unique and distinctive styling did not hide its mechanical failings, and a recall programme even had to be initiated to sort out its ills.

It was decided to replace all the British and American models with a brand-new range that would be uniform around the world, except for local features and requirements. Designed under the 6X designation, the tractors would be launched towards the end of 1964 as the 1000 Series and would be built at the Highland Park factory in the USA, Antwerp in Belgium and also at a brand-new factory in Basildon, Essex.

A lot of time and money had been spent on developing what was a brand-new range of four tractors, but it was perhaps asking a bit too much for everything to go smoothly when brand-new designs were introduced at the same time as a brand-new factory and there were teething problems, particularly to begin with.

In the UK the tractors not only bore their new number identities, but also the name of the model they largely replaced. Therefore the smallest, the 37 hp 2000 was also known as the Dexta, the 46 hp 3000 the Super Dexta, the 55 hp 4000 the Major and the 65 hp 5000 the Super Major. New Ford three-cylinder engines powered the 2000, 3000 and 4000, while the 5000 was a four-cylinder tractor. In America a revised version of the 6000, known as the Commander, was sold as part of the new range.

Smallest of the new 100 Series, or 6X models, was the little 37 hp 2000, which is shown here dwarfed by the Weeks grain trailer it is pulling.

43

Next up in size was the 3000 and these and the larger models had white front grilles with full-length white bonnet stripes, as opposed to the 2000 with its almost all-over blue.

This early 4000 shows the Major name next to the steering wheel and is a largely original Lincolnshire example.

A restored 5000, which, as the flagship model in the new range, was the only four-cylinder model. It originally produced 65 hp before being increased to 67 hp when it failed to perform as well as expected.

In North America the 6000 was replaced by the reworked 6000 Commander, which featured tinwork to give it a family resemblance to the new 1000 Series. This example is using a set of Ransomes disc harrows in Wiltshire, but Ford never sold these tractors in the UK when they were in production.

Brand-new styling, as befitted a completely new design, gave the four new models a very modern look, with a distinctive front cowl treatment and slats down the sides of the bonnets. Transmission options included six- and eight-speed manual gearboxes plus the Select-O-Speed powershift, which was now offered for the first time outside of the USA.

Various issues were encountered with the new tractors and the 5000 was often criticised for lacking power, the result seeing the engine soon being tweaked to produce 67 hp. But it was the Select-O-Speed transmission that was most often disliked in the UK, with drivers not being able to quite get to grips with its smooth operation. The result was that it was soon nicknamed the Jerk–O-Matic!

Although most of the problems of the 6X range were gradually addressed, the major update of 1968 saw them all finally put to rest for good. Known as the Ford Force range, this upgrade saw less fussy bonnet styling throughout the range from the 2000 to the 5000 and the flagship model now boasted 75 hp.

1968 also saw the 6000 model finally being replaced by something better as the 115 hp six-cylinder 8000 became the most powerful Ford tractor to be built so far. A brand-new engine soon proved to dispel any memories of the problems of the 6000 model and in 1969 a 145 hp turbocharged version appeared called the 9000, taking the power available even higher!

The Ford Force range of 1968 saw the small issues with the original 6X range ironed out, as well as a much less fussy style of tinwork, as shown on this Force 3000.

Right: Narrow versions of the Force 3000 were also built, as demonstrated by this extremely narrow tractor, which were ideal for work in vineyards and orchards.

Below: Ford offered a weather cab for fitting to the Force range and used canvas cladding over a metal frame and a glass windshield. The distinctive design was ideal for protecting the driver from the weather but was not the easiest to get in and out of.

Left: Ford returned to the six-cylinder concept with the 8000 of 1968. A new 401 cubic inch diesel engine was used to great success and this 115 hp tractor proved much better than the earlier 6000 model and was only built in the USA

Below: In Europe many American-built 8000 tractors were equipped with a powered front axle to provide a very efficient four-wheel drive tractor, which was usually fitted by the Ford dealers themselves.

A compact tractor was added to the Force line-up in the USA in 1970 with the introduction of the 1000. This 23 hp tractor was built in Japan by Shibaura and was rebadged as a Ford. This tractor was the forerunner of a whole range of later compact machines, which were also then sold in the UK and Europe through to the 1990s.

1970 also saw the introduction of safety cab legislation in the UK, and as a result all new tractors had to be fitted with a safety cab that was approved and could be relied on to protect the driver in the event of a roll-over accident. Ford turned to its industrial division in Copenhagen, who came up with a very stylish cab that was then fitted to the Force range. The two smaller tractors had a version with a soft-top vinyl roof while the 4000 and 5000 had a hard-topped version with flat rear mudguards.

At 75 hp the 5000 was a large tractor, but was not big enough for some. Many British farmers were reluctant to adopt the much heavier six-cylinder 8000 and so Ford came up with the answer when they introduced the 7000. This tractor, basically an upgraded and strengthened 5000, featured a turbocharger to boost power up to 94 hp and in the process produce a lightweight but very powerful machine. At the same time a new hydraulic system, known as Load Monitor, was introduced to the 7000 and 5000 and allowed for better hydraulic control of trailed implements such as long, heavy ploughs.

The 126 hp 8600 and 145 hp 9600 replaced the earlier 8000 and 9000 and were introduced ahead of the rest of the 600 Series. Now built in Belgium as well as in the USA, a Hara cab was fitted to meet UK legislation.

A Dual Power transmission also became available on the bigger Ford tractors, this being a gear splitter giving a high and low ratio in each standard gear, making the standard 8 x 2 box now a 16 x 4 unit.

The little 1000 was not really a Ford, being built as it was in Japan by Shibaura. It was the first Ford-badged compact tractor and would give rise to a whole range, all of which were supplied by the Japanese firm.

Safety cab legislation resulted in tractor manufacturers having to fit cabs with roll-over protection to all new machines and there was also an after-market need, catered for by the likes of Lambourn, who made the cab fitted to this Force 4000 in East Sussex, which used flexible cladding at the sides and rear.

Ford brought out a specially designed safety cab for the Force range, as shown by this Force 3000. The smaller tractors kept their round rear mudguards.

This Force 4000 has the Ford safety cab and flat mudguards designed especially to accommodate the new cab.

A 75 hp Ford Force 5000 and a Ransomes plough at work in Essex.

Cultivating in Wiltshire with a Ford Force 5000 fitted with a full set of front weights.

The Ford 7000 saw turbocharging make a real impact on British farming, with its turbocharged four-cylinder engine producing 94 hp and the tall air intake cleaner, with its pre-cleaner bowl, making these tractors look very impressive.

The 126 hp 8600 replaced the 8000 and was now built in Belgium, at Antwerp, as well as in the USA.

The 9600 put out 145 hp from its turbocharged six-cylinder engine. This example is fitted with the North American style rear mudguards and a roll-over frame.

A Ford 8600 with the Swedish-sourced Hara safety cab that was fitted to examples sold in the UK.

The safety cab had certainly prevented some serious injuries by protecting the operator, but had caused further problems due to the fact that they amplified the noise the driver was exposed to. Legislation came into force during 1976 that required manufacturers to fit an approved quiet cab and reduce the noise level at the driver's ear. Ford were developing a new quiet cab to use across the tractor range, but it was not yet ready for production in 1975 when the replacement for the Force range was unveiled. Known as the 7A1 range, or more usually as the 600 Series, this new line-up was based heavily on the Force range but with tweaked engines, and also two new models in the form of the 52 hp 4100 and 68 hp 5600, the 78 hp 6600 being an evolution of the original 5000. They did not look very different as, until the spring of 1976, they still wore the same safety cab as the earlier range.

The arrival of the new Q cab changed the look of the 600 Series completely and gained the internal designation of 7A2. The stylish new cab not only kept noise levels inside very low, but also looked extremely modern, its curves and rounded edges soon giving it the nickname of the 'bubble' cab.

The new cab greatly added to driver comfort, but the new 700 Series added to that even further by providing a flat floor cab thanks to their higher build. The 6700 and 7700 were luxury versions of the 6600 and 7600, while the 128 hp 8700 and 153 hp 9700 replaced the larger six-cylinder tractors, these last two being the first Ford tractors to also have the option of factory-fitted four-wheel drive.

The new 600 Series saw a couple of new models introduced compared with the Force range they replaced. The 4100 was something of a hybrid tractor, using both 3000 and 4000 components. This one has been fitted with a later engine block after the original went porous – a common Ford fault.

The new 4600 replaced the 4000 and now had 62 hp on tap from its three-cylinder engine. Cage wheels are fitted to this Norfolk example for drilling cereals with a Nordsten drill.

The 5600 was the other new model introduced in 1975 and was a 68 hp machine. Dual wheels help with traction while breaking down ploughed land in Lincolnshire with a power harrow.

The 78 hp 6600 replaced the 5000 and would go on to be a very popular model in the range. (Photograph: Kim Parks)

With the new quiet cab fitted, this 4600 is turning hay in Suffolk.

This 6600 has been fitted with a tall air cleaner stalk and pre-cleaner bowl on the bonnet, which was not a standard feature except for some export markets.

The 7600 was usually fitted with the tall air intake though, as it was thought to be important that the turbocharged engine should be protected from dust and dirt. Based on a livestock farm in Sussex, this tractor's sole job is to chop and spread straw for bedding.

The 7700 was the deluxe version of the 7600 in the high-bonnet 700 Series. This is a late example, fitted with the more powerful 103 hp PowerPlus engine that was introduced in 1980.

The 9700 was the largest tractor in the Ford range when introduced in 1976 as well as the flagship of the 700 Series. With 153 hp available, even a two-wheel drive version was capable of pulling a nine-furrow conventional plough.

Most 9700 tractors sold in the UK were fitted with the ZF front drive axle from new and were built in Belgium. Four-wheel drive gave these tractors even more pulling power.

Left: The view into the flat-floor Q cab fitted to the 9700 tractor. The quiet cab cost Ford a lot of money to develop, but the end result was an extremely good cab and a nice place to spend a long working day in.

Below: The dashboard fitted to the 700 Series was clear and well laid out under the steering wheel.

An Economy version of the 7600 was produced for those who wanted a simpler and cheaper machine.

An odd tractor arrived in 1978 and it was something of an orphan. The 8100 was not even built at Basildon, being assembled at Fleet in Hampshire by County Commercial Cars, a company that built four-wheel drive versions of Ford tractors. The 8100 was offered as a two-wheel drive tractor with 115 hp from a six-cylinder Ford truck engine and was designed to answer a demand for higher power take-off horsepower for driving power-hungry implements such as forage harvesters. Four-wheel drive versions were later offered and County also built an improved 8200 version.

Power was something that was required more and more by the world's farmers as equipment got ever larger and wider. Ford's largest tractor was the 153 hp 9700 in 1977, but, by getting the Steiger company of North Dakota in the USA to build badge-engineered versions of its four-wheel drive, articulated steer monster tractors, Ford could now offer machines of over 300 hp. The FW series was a four-model range, only two of which were imported into the UK; the 295 hp FW-30 and 335 hp FW-60, both powered by Cummins engines.

The TW range followed this in 1979 and saw the 8700 become the 128 hp TW-10, the 9700 the 153 hp TW-20 and a new flagship model introduced. The TW-30 was certainly a big machine, based on the same engine as the smaller models; the TW-30 featured a turbocharger and an intercooler to produce 188 hp. This was a big machine and the extra-long bonnet also concealed a front-mounted additional fuel tank, and made it look even larger. TW, by the way, actually stood for two-wheel drive as opposed to the FW four-wheel drive tractors, but ironically most of the TW tractors sold in the UK were of four-wheel drive layout – the opposite to the position in the USA where this size of tractor was traditionally of two-wheel drive layout.

The 8100 was assembled for Ford by County Commercial Cars in their factory in Hampshire. Designed for those who wanted more power at the power take-off, the 8100 used a 115 hp six-cylinder engine in the four-cylinder chassis and needed a frame to support the unstressed engine block. (Photograph: Kim Parks)

Steiger built the FW Series for Ford, but only the FW-30 and FW-60 were imported into the UK. A Cummins six-cylinder engine provided the power in the FW-30, which was rated the same as the American spec FW-40 model.

The FW-60 was an impressive machine with over 300 hp on tap, and was one of the largest tractors available to British farmers at the end of the 1970s.

The TW tractors were built at a factory in Romeo in the USA and in Antwerp in Belgium for various world markets. For Australia, special versions of the USA and Basildon-built tractors were produced, including, from 1980, the 8401 model, which was based on a 7700 tractor but with the six-cylinder engine from a TW-10 and with very distinctive styling and Australian-sourced cabs. For a time, in the mid to late 1980s, Ford also bought articulated tractors from the Australian Waltanna company, including the FW-25 and FW-35, which were based on TW Series skid units.

A replacement for the 9700 came in 1979 with the launch of the TW range and the TW-20 model. It kept the same 153 hp rating as the older model and this one is at work cultivating in Wiltshire. (Photograph: Kay Balls)

The TW-30 was a new model featuring 188 hp and a long bonnet that was designed to incorporate a second fuel tank at the front. The side panels slid forwards for access to the engine. The largest Ford model proper, the TW-30, was a very imposing beast, and this one is using a Ransomes subsoiler in Norfolk.

Chapter 7

Perfect 10

With new high-horsepower models unveiled, attention moved to the rest of the range in 1981 when all the three- and four-cylinder models were replaced by the new Series 10 tractors. From the new 44 hp 2610 up to the 98 hp 7710, all these new models featured a revised bonnet design incorporating a lift-up nose cone, except on the high-bonnet 6710 and 7710. The Q cab was still fitted but the main advance was the fitting of a synchromesh transmission in a Ford tractor for the first time. This gave a much smoother operation of the gears and allowed for on-the-move shifting. The new SynchroShift transmission was well designed except for one issue – the operation! The bigger tractors were fine with their floor-mounted levers, but on all the models from the 103 hp 7610 down, a new column shift was introduced, which was far from logical in operation and it was only too easy to select the wrong gear!

Although it was possible to get used to the operation of the gearshift, it was pretty much universally hated, and was called the 'Rubik's Cube' box by many due to it being so hard to find the desired ratio. It was eventually changed for a more conventional 'H' pattern shift, still being mounted on the steering column.

A bigger Series 10 model arrived in 1982 when the 116 hp 8210 replaced the 8100 and the 8200, and was finally built at Basildon. Further new models arrived in 1984, the 2910 and 3910 replacing the earlier models with stronger back-ends while the 7910 combined the same engine as used in the 8210 but de-rated to 103 hp.

The TW range was updated in 1983 to become the 143 hp TW-15, 163 hp TW-25 and TW-35, with power being increased across the range, meaning the TW-35 now put out 195 hp. The TW-25 also now gained the longer bonnet of the flagship model and the extra front-mounted fuel tank. Another smaller model was also introduced called the TW-5, with 128 hp available, but this was never imported into the UK.

Force II was a major upgrade of the Series 10 and TW models in 1985. The big news was the new Super Q cab, with its lower height, improved noise deadening, extra worklights and the fitting of a flat floor throughout the range teamed with floor-mounted gear levers to finally do away with the column shift design that had been so disliked. The Super Q cab was only offered on tractors from the 5610 upwards, with the smaller three-cylinder machines having the Sekura-built LP or AP cabs fitted, which were first introduced a year earlier. These cabs were also offered on the 72 hp 5610, 86 hp 6610 and 103 hp 7610 as an alternative to the Super Q.

The 6610 replaced the 6600 in 1981 and this example is in the middle of planting sugar beet in Suffolk.

Fitted with the SynchroShift synchromesh transmission, the 6610 was a popular member of the new Series 10 range.

In the cab of a 4610 we can see the two gear levers positioned under the steering wheel, which are used to control the SynchroShift transmission.

The largest of the Series 10 models was the 116 hp 8210, which arrived a year later, in 1982. It was originally the only six-cylinder tractor in the range.

The 335 was part of the highway range of industrial Series 10 models often used by local municipal authorities.

Arriving in 1984, the 7910 was a late addition to the first generation of Series 10 tractors and was a six-cylinder, 103 hp tractor. Most were of four-wheel drive configuration, making this two-wheel drive machine turning grass in Lancashire quite a rare beast.

Inside the Q cab of the 7910, the floor-mounted gear levers and the hydraulic controls are clearly visible.

The TW range received a face-lift in 1983 and this included a boost in power output. The smallest offered in Britain was the TW-15, with 143 hp on tap.

A longer bonnet complete with an extra front-mounted fuel tank was a feature of the new TW-25 model. In Britain, TW tractors were usually of four-wheel drive format, while in the USA they were almost all two-wheel drive.

New flagship of the range was the TW-35 with 195 hp now on offer under that long bonnet. This one is using a subsoiler in Wiltshire.

From 1985 the Force II Series 10 range replaced the original models. The 4610 Force II model was not offered with the new Super Q cab but many chose the AP version over the low-profile LP cabin.

The Sekura-built AP cab gave good all-round visibility and was certainly comfortable, if rather more basic than the top-of-the-range Super Q.

72 hp was the power offered by the 5610, which was the smallest four-cylinder tractor in the Force II line-up.

Many livestock farms took advantage of the lower profile of the tractors fitted with the AP and LP cabs, especially for loader work. Even with all the opening windows, the doors on this one have been removed during a hot summer.

The Super Q cab was a superb update of the original Q cab but with a flat floor and no signs of a column-mounted gearshift. This Force II 7610 is rowing up grass in Lancashire.

Two-wheel drive versions of the Series 10 tractors were still quite popular, even when used with a front loader, as demonstrated by this 7610 disc harrowing in Lincolnshire.

Right: The Force II tractors with Super Q cab looked superb, as shown by this as new-looking Force II 7610.

Below: The 7710 Force II was the sole four-cylinder representative of what had been the 700 Series. Some were equipped with a tall air cleaner intake, as on this Wiltshire example.

The 8210 carried over into Force II guise without any significant changes; both this and the 7710 were already fitted with floor-mounted gear levers and a flat-floor cab so were not greatly changed in 1985.

Under the Force II upgrade, the TW range also benefitted from the Super Q cab, as shown by this TW-15 Force II. (Photograph: Kim Parks)

The white front grilles and white rear mudguards introduced under the Force II programme suited the larger two TW models well. This TW-25 Force II is performing well with a large conventional plough in Wiltshire. (Photograph: Kim Parks)

A Ford TW-35 Force II looking the part at a working day in Wiltshire, complete with subsoiler. (Photo: Kay Balls)

This little 3910 Force II shows off the LP cab as it turns a crop of hay in Suffolk.

1985 also saw the Ford Motor Company buy the American New Holland farm equipment manufacturer from its parent, Sperry Rand. New Holland was a force in grassland equipment as well as balers and combine harvesters and had a strong European presence thanks to its takeover of the Belgian Claeys business. The merger of the Ford tractor operations and New Holland saw a new company formed called Ford New Holland.

In 1987 a new model was introduced and it would prove to become the best-selling of the whole Ford tractor range. The 7810 featured the six-cylinder engine used in the larger models but mated it to the lighter 7610 chassis, giving a lightweight but powerful 103 hp tractor. Designed by South Essex Motors for Ford, this light 100 hp tractor would prove to be just what the market wanted.

1987 was also the year that Ford New Holland bought the Canadian Versatile firm with its range of grain handling equipment and four-wheel drive articulated tractors. The FW range had fizzled out and so some of the Versatile models were soon offered for sale in Ford colours and branding.

In 1989 the Generation III Series 10 models were the result of a further upgrade with added power and also new models, including the 84 hp 6410 and 94 hp 6810.

The 8210 now received an ElectroShift electronically controlled hydraulic system – a first for a Ford tractor. This system gave electronic draft control and was adjustable by using a simple control knob, and also included a digital readout. Electronics were being adopted by many of the major tractor manufacturers and soon such items would become the norm.

The last of the Force II tractors to appear was the 7810 in 1987. With its six-cylinder engine producing 103 hp, it soon became a best-seller, offering more power in a lighter package.

The Generation III Series 10 tractors launched in 1989 included two new models, the smallest of which was the 84 hp 6410.

The other new model was the 94 hp 6810 Generation III. This was basically a de-rated 7610 and still featured a turbocharger.

The three-cylinder members of the Series 10 family were also briefly part of the Generation III family, with the 4610 still being offered with LP or AP cab options.

A very original Ford 6810 in a barn in Lancashire while taking a break from carting grass for silage.

Above: The popularity of the 7810 increased even more when in Generation III guise and was used for a huge variety of tasks, including ploughing.

Right: The view into the Super Q cab of a 7810 Generation III, showing the flat floor to good effect.

With the introduction of the Generation III 8210 model, a new era of electronic implement control on a Ford tractor was born thanks to it being equipped with ElectroLink hydraulics.

The ElectroLink control terminal in the 8210 was placed neatly to the right of the main controls and included a digital readout.

To mark twenty-five years of tractor production at the Basildon factory, a special batch of silver-painted Ford 7810 tractors were produced. Almost ignored at the time, they fetch high prices today from collectors.

Chapter 8

Futuristic Features

The TW range also disappeared in 1989, although the three models remained much as before under the guise of the new 30 Series. Once again, new sophistication was at the heart of the new 8630, 8730 and 8830, which, outwardly, looked identical to the earlier TW range with the same power outputs, but now wore black grilles and black rear mudguards. Inside, though, the new tractors were offered with a full powershift transmission controlled by a single stubby lever up to the right of the driver, which also controlled the direction of travel. A digital display was also fitted to show the gear selected. Dual Power versions of the 30 Series were also built for those who did not require the new transmission technology.

A full powershift transmission was at the heart of the Series 30 models, which replaced the TW range in 1989. The smallest in Britain was the 8630, although a smaller 8530 was also built for certain markets.

The digital dashboard of the Series 30, as well as the Generation III tractors when fitted, included a performance monitor.

The 8730 replaced the TW-25 and retained the long bonnet. Front-mounted push-pull ploughs were an innovation of the 1980s and were a good way to harness all the power of a tractor – in this case, 163 hp.

The biggest was the 195 hp 8830, the full powershift gearbox proving ideal for high draft work. Extra front weights on an extended carrier have been fitted to this one, as well as belly weights between the front wheels.

The stubby gear selector for the powershift transmission was positioned up next to the hydraulic quadrant and included a digital readout of the gear selected. Both speed and travel direction are controlled by this same lever.

The 8830 is equally at home on high-power requirements, such as using a Mengele trailed forage harvester.

A smaller 30 Series appeared in 1990, using the black design cues of the six-cylinder tractors but applying it to the much smaller three-cylinder models. Basically replacing the Series 10 models from the 4610 down, these new tractors included a new cab and a new shuttle gearbox, and were later joined by the 73 hp four-cylinder 5030.

All these developments were gradually modernising a tractor line that could trace its evolution from 1964, which is why Ford had already been working on replacements for the Series 10 and Series 30 models for some time. Circumstances were to overtake these plans, however, when it was announced that Ford New Holland was to merge with Fiatagri. In truth this was a takeover, and the new owners would be the Italian Fiat company – the deal being finalised in 1991.

In many ways the Ford and Fiatagri product lines complemented each other quite well, but it would only be a matter of time before both would be merged together. For now though, it was more or less business as usual, and this meant the launch of the all-new Ford Series 40 in 1991. Replacing what was left of the Series 10 Generation III tractors, the Series 40 brought a 'new from the ground up' range of tractors featuring new cabs, new PowerStar engines and new electronically controlled features including the ElectroLink hydraulic system and the ElectroShift semi-powershift transmission available on the top-of-the-range SLE models, while the SL tractors were of lower specification and included a manual transmission that was later equipped with Dual Power gear splitter.

The range began with the 75 hp four-cylinder 5640 and went up to the six-cylinder 120 hp 8340, and the new slightly sloping bonnets with slim exhaust and under bonnet air cleaners made them look very modern indeed.

The 8340 was later taken up to 125 hp and the range, built in Basildon, was very well received and was heralded as the first new Ford tractor for a generation.

The 4630 replaced the 4610 model in the new small Series 30 range, which incorporated a new shuttle gearbox and a new cab. This example is all set for cultivation work with duals fitted all round.

Series 40 was a major upgrade to the Series 10 line, with a whole host of new features making these completely new machines. The 6640 was the second smallest and shows off the new Super Lux cab, the sloping bonnet with front-mounted fuel tank and also the plastic engine side coverings.

The 7740 was the smallest turbocharged model with 95 hp under its bonnet. This is a top-spec SLE model, with semi-powershift transmission and electronically controlled rear linkage.

SL versions of the 7740 were equipped with manual hydraulic controls and a manual transmission, which was not originally fitted with Dual Power.

Above left: The view into the SuperLux cab of a 7740 SL, showing the wide opening door and roof hatch.

Above right: The SL version of the Ford 7810 SL, which is also not fitted with plastic engine side covers and there are no work lights on the front of the cab roof.

Below: The controls for the ElectroShift semi-powershift transmission in the 7840 SLE.

A Ford 7840 SLE drilling spring barley with a power harrow and drill combination as well as cage wheels – a rare sight on a 1990s tractor.

The biggest of the Series 40 models was the 8340, which ended up being a 125 hp machine. This took the Basildon-built tractor range up into new higher-horsepower territory.

Chapter 9

The Name Disappears

When Fiat took over Ford New Holland, Ford allowed the use of the Ford name and famous oval trademark for only a few years. Gradually, both the Ford and Fiatagri tractor lines were brought together under the new brand name of New Holland.

Before then though, the very last tractor to have input from Ford was launched. The Series 70 were high-horsepower tractors designed to replace the big Series 30 models based on the old TW platform. The only thing in common with their replacements was the Funk-sourced full powershift transmission incorporated into the tractor – the rest was

The baby of the new Series 70 was the 8670, with 170 hp under its sleek new bonnet. These were impressive ultra-modern tractors when introduced in 1994, with many novel features. (Photograph: Sascha Jussen)

all-new. Built in the Versatile factory in Canada, the 70 Series was made up of four models, from the 180 hp 8670 to the 240 hp 8970, and incorporated many novel features, including the option of a SuperSteer front axle that pivoted, giving an unheard of sixty-five degree turning angle for tight headland turns. No less than three on-board computers were used to run the machine and the new cab included a moveable console that housed the gear lever and other controls.

With a large sloping bonnet that could be tilted upwards for maintenance to the engine, these were really modern-looking and sophisticated prime movers, which were built ready for twenty-first-century farming. Another first was that the 70 Series was also available through Fiatagri dealers as the G Series in terracotta paint – this being the first new model to be offered in such a way.

By the mid-1990s the Ford and Fiatagri names were getting smaller and the Ford oval badge was replaced by a new blue version of the leaf emblem used by Fiatagri. The New Holland name was now prominent and the 40 Series had been given a revamp with new grey chassis and New Holland branding, plus a blue roof to the cab. The 7840 would remain the last of the line, surviving in production up to 1998 – two years after the other models had been discontinued. They were replaced by the New Holland TS range, which did not carry the Ford name. It was the end of an era, and it is perhaps fitting that the last Basildon-built tractor with Ford heritage was the best-selling 7840 model.

The Fiatagri version of the Series 70 was the G Series with the G240 being the equivalent to the 240 hp 8970 and exactly the same except for the colour scheme.

89

The stubby gear lever used in the Series 30 was retained but modified to fit into a right-hand side console with the hydraulic and other controls.

The Ford name was slowly made smaller as the New Holland name took precedence following the Fiatagri and Ford New Holland merger of 1991.

Series 40 tractors evolved over the first half of the 1990s and were gradually brought into line with the New Holland ranges being introduced by the middle of the decade. This 7840 SLE shows the first signs of change, with a blue roof to the cab and a grey chassis.

A Ford 6640 also showing the first changes of identity with its blue cab roof while rolling cultivated land in Suffolk.

Next, the Ford name lost its traditional place in large letters on the bonnet to the words 'New Holland'. The Ford name was still kept but only in small letters under the model number, as on this two-wheel drive 6640.

A New Holland 7740 rowing up grass complete with a Howard front loader.

A blue version of the terracotta Fiatagri leaf emblem was adopted as the logo for New Holland and was used on the front of the Series 40 tractors in place of the familiar Ford oval.

One of the very last 7840 tractors built is at work in East Sussex, showing how the New Holland name completely replaced that of Ford by the end of production on these, the last completely Ford-designed tractors.

The Canadian-built Series 70 tractors survived longer than the Series 40 and evolved into an improved range in 2001, as shown by this 8970A, which is working down a seedbed in Wiltshire. By then the factory in which they were being built and their design rights had been sold to Bühler, who carried on making them for New Holland until 2002.

Chapter 10

Legacy

Ford tractors changed the world! That is quite a statement, but it is also a very true one. Without the brilliance of Henry Ford's mass-production techniques, and without his determination to produce a farm tractor even when his company did not want to, the history of farm mechanisation might have been very different.

During the First World War, his first tractors helped Britain survive. The Model N did the same during the Second World War and the Major rebuilt agriculture afterwards. The New Major and the Dexta became farming icons of the 1950s and '60s and the 1000 Series set standards in farm tractors that were not surpassed for several decades.

Earlier generations of Ford tractors are well represented in the preservation movement, as evidenced by this impressive line-up of Fordson Major tractors at a show.

Not only did Ford tractors shape the way people farmed, they were also used by many small companies to form the basis of other machines both for agriculture and industry. From the days of the Model F through to the Series 10, companies in the USA, in Europe and the UK took skid units of Ford tractors, consisting of engine and transmission, and adapted them to a multitude of different uses. In Britain the best remembered of these firms are County Commercial Cars, Roadless Traction, Muir-Hill and Doe, all of whom specialised in four-wheel drive versions of Ford tractors.

Today the legacy of the Ford tractor lives on despite the name disappearing from tractors in the 1990s. New Holland continues to be one of the world's leading tractor and farm machinery companies and still assembles tractors in the factory at Basildon in Essex. In 1999 Fiat bought Case IH and merged the two businesses to form CNH and New Holland is today part of the CNH Industrial Group.

There are also many, many thousands of Ford tractors still working on farms every day, as well as many more in private collections. Enthusiast clubs, such as the Ford & Fordson Association and Blue Force in the UK, keep the Ford flame alive and help members to preserve their machines.

One thing is for sure; the Ford tractor has made such an impression on so many people around the world that it will never be easily forgotten. And that is how it should be!

Ford tractors are also still used in farming, as shown by this 8830 and Silver Jubilee 7810, which are collecting grass for silage in Lancashire.

The Basildon factory set up by Ford in 1964 still produces New Holland tractors, including the T7.315 of over 300 hp, which is shown here in the Blue Power special paint finish.